A GUIDE TO FOOD ADDITIVES

By
Donald and Elisabeth Saulson

VPS Publishing
Huntington Beach, California

VPS Publishing
P.O. Box 2705-261
Huntington Beach, CA 92649

Cataloging in Publication Data

        Saulson, Donald, 1937-
        Saulson, Elisabeth, 1961-
        A pocket guide to food additives.
        Includes bibliographical references
        I. Food additives I. Title
Library of Congress Catalog Card Number 91-90996
ISBN 0-9629606-0-8

First Printing, June 1991
Printed in the United States of America

A POCKET GUIDE TO FOOD ADDITIVES

CHAPTER 1. INTRODUCTION

We have been living through an era of "convenience foods." It's time we asked ourselves who are these foods really convenient for - us or those that sell them.

Food additives are used to make a product look, taste and stay saleable for as long as possible. If you purchase a loaf of white bread loaded with chemical additives, chances are the bread will be fresh for a relatively long time; but at what cost do we really want things to stay fresh? Is it really worth stripping the bread of most of its nutritional value and then adding insult to injury by adding such chemicals as bleach, ammonium, and peroxide.

On the other hand, if you buy a loaf of fresh whole wheat bread with no chemical additives, you can be sure you're getting a nutritional product. However, it must be sold to you fairly quickly from the time it was baked fresh. Products with no chemical additives have a short shelf life since they usually do not contain ingredients that prevent them from getting stale, growing mold, etc...

Many health advocates have opposed the use of food additives because many of the chemicals involved have been linked to degenerative diseases such as cancer. Would we be a happier and healthier society without chemical additives in our diet? The consumer must try to understand the risks and weigh them against the benefits. A chemical preservative may be linked to higher incidence of cancer, but the link may be uncertain or only valid for excessive doses; while <u>without</u> the preservative the consumer might risk exposure to a deadly bacteria. Also the consumer will recognize that in many cases an additive introduces nutrients that replace those lost in processing or even fortify it above its original strength. Probably the best advice to give a consumer about chemical additives is to research each additive you're not sure about, weigh the pros and cons, and go with what feels right for you. This guide can help you do just that.

CHAPTER 2. SPECIFIC USES OF FOOD ADDITIVES

Additives are used to control the food's appearance, taste, freshness, and nutrition. It may also assist in food processing. This chapter identifies and defines the terms used in industry for those functions.

COLOR

Bleaching Agents: A whitener. Accelerates the natural aging process of flour (improves baking).
Coloring: A coloring dye. Improves appearance.
Color Fixatives: Keeps the natural color fixed.

TASTE

Acidulant: It is an acidifier that adds a tangy or tart taste to a product.
Flavor Carrier or Flavor Dispersant: It dissolves a flavor and disperses it throughout the product.
Flavor Enhancer: Improves or brings out the natural flavor.
Flavoring: Adds a particular flavor to a product.
Salt Substitute: A substitute for salt that does not contain sodium.
Seasoning: It makes food more savory or flavorful.
Sweetener: It gives food a sugary taste. **Artificial Sweeteners** are synthetically produced, and contains much fewer calories than sugar. **Natural Sweeteners** are sugars found in plants.

ACIDITY/ALKALINITY

Acidifier and Acidulant: Increases the products acidity or reduces alkalinity.
Alkalizer: It reduces acidity or increases alkalinity.
Buffer: It helps regulate the pH (acidity or alkalinity) to a prescribed value.
Neutralizer: A buffer that eliminates both acidity and alkalinity, keeping the product neutral.

MOISTURE CONTROL

Drying Agent: Decreases a product's moisture.
Humectant: Keeps product from drying out.
Surfactant or Wetting Agent: Keeps product moist.

PRESERVATION

Acidifier: Increasing acidity also acts as a preservative (affects microbe growth) and act as an antibrowning agent.
Antibrowning: Prevents browning due to oxidation.
Anticrystallization: It prevents the forming of crystals which can cloud up certain products.
Antifungal: It inhibits the growth of rope, mold, and fungus.
Antimicrobial: It inhibits bacteria growth.
Antioxidant: It slows down the spoiling of fats due to oxidation.
Firming Agent: It maintains the firmness and texture of a product.

Preservative: Prevents spoiling due to oxidation and growth of fungus and bacteria.

Sequestrant: It binds and precipitates out metal ions that would affect the products appearance, flavor and texture.

PROCESSING AIDS

Anticaking Agent: Prevents powders from hardening.

Binder: Used for holding substances together.

Bread Improver: Oxidizing substances which mature or "improve" the quality of bread.

Catalyst: A substance which brings about or hastens a reaction, but is not consumed or changed in the reaction.

Clarifier: Used for removing small particles from liquids.

Defoaming (or Antifoaming) Agent: Prevents or lessens the foam in liquids.

Dough Conditioners: For making dough easier to manage.

Emulsifier: It stabilizes mixtures (like oil and water) and keeps them at a uniform consistency.

Fermenting Agent: It breaks down compounds into alcohol and carbon dioxide.

Filler: Used for replacing certain ingredients with less costly substitutes.

Leavener: For lightening the texture by releasing gases.

Maturing Agent: Speeds up the aging process in order to make more manageable dough.

Refining Agent: Purifies products by eliminating trace elements.

Solidifier: Helps to solidify the product.

Solvent: A liquid that can dissolve one or more substances.

Stabilizer: Used to keep a uniform consistency.
Texturizer: Used to improve texture.
Thickener: For adding body and thickness to the texture.
Yeast Food: It speeds up the fermentation process.

NUTRITIONAL

Chemicals used to enrich or fortify foods by adding important nutrients such as vitamins, minerals, essential amino acids and fatty acids, and other food value.

CHAPTER 3. HOW TO USE THIS GUIDE

BACKGROUND. The effects of food additives on humans has been evaluated in two ways. Many of the ingredients have had widespread use for many years before regulation began (see section on <u>The GRAS List</u> at the end of this chapter). From observation of the health of consumers correlated with the substances they consumed, some conclusions can be drawn regarding the safety of various ingredients. The problem is that this is not very precise, and health effects may not show up for years, or be masked by the effects of other substances. An attempt to correct this, was to use massive doses on laboratory animals and evaluate the animals for toxic and mutagenic effects. However, many investigators have now concluded that those attempts are seriously flawed. The correlation between effects on laboratory animals and humans is too weak, and the extrapolation of a safe dose from a massive dose, is too inaccurate. New approaches need to be developed. Nevertheless, researchers have combed the data that is available and have rated the ingredients with respect to their relative safety. This was done by Fredberg and Gortner in reference (1) and by Jacobson in reference (2). The references are listed in Appendix I. Their recommendations were used to develop the safety ratings that appear in the Additives Dictionary chapter of this guide.

HOW TO READ THE DICTIONARY.

SAFETY RATING. The dictionary lists the food additive chemicals in alphabetical order. After the chemical's name appears safety rating codes in brackets []. The codes are shown in Table I. These ratings are based on toxicity effects at the highest levels of concentration expected in most foods. Additives that could be toxic in very high concentrations can still be safe for their intended use. Sugars and saturated fats are sometimes given a "cautionary" or even "avoid" rating in some references. Evidence is that for general health and to minimize heart disease, these foods should only comprise a small percentage of one's diet (particularly if you are not engaging in much aerobic exercise.)

PRIMARY USES. After the brackets with the safety rating code, in parenthesis () is an abbreviated list of the primary uses of the additive. For example, (Anticaking/ Neutralizer) indicates that the additive is an Anticaking Agent and acts as a neutralizer to eliminate acidity or alkalinity. The definitions of these terms can be found in Chapter 2. Specific Uses of Food Additives.

OTHER INFORMATION. After the parenthesis is a brief discussion of the source or composition of the substance. The user of this guide will be referring to the label of a particular food (see section in this chapter about <u>Facts about Food Labels</u>). The additive will be in that food item.

Therefore, it is not necessary for this guide to list the types of foods where the additive is usually used. That information can be found in the references in Appendix I.

HARMFUL EFFECTS. For some additives a brief description of major harmful effects that have been observed on humans will appear in brackets. Those reported for laboratory animals are not listed since the results are controversial.

THE GRAS LIST. Since there is no substance which is 100% safe for 100% of the population, the FDA can not claim any food additive to be completely safe. With this in mind, the GRAS list was established in 1958. GRAS stands for Generally Recognized As Safe. With the exception of "good manu-facturing practices" most food additives on the GRAS list may be employed without any limits. Unfortunately, being on the GRAS list is no guarantee that a food additive is even relatively safe. Many additives were previously on the GRAS list, but later taken off and sometimes banned altogether when tests proved them to be dangerous. Some examples of this are cyclamates, saccharin and certain artificial colors. If at the end of the entry, the letters GRAS appear, it signifies that the substance is on the GRAS list.

THE GLOSSARY. Appendix II. is a glossary that will help define some of the terms found in the dictionary.

THE QUICK GUIDE

In Appendix III is a quick guide to food additives. It groups together for quick reference those substances in the dictionary that are listed AVOID and those that are listed CAUTION.

FACTS ABOUT FOOD LABELS

There are laws and guidelines involved in the making of a food label. A label must include (in a language understood by the consumer):

> * The name of the manufacturer.
> * The name of the packer or distributer.
> * The ingredients, in order of quantity.
> * The volume or net weight of contents.
> * Nutritional labeling, IF any nutritional claims are made on the label.

A food label need not mention:

> * Any artificial colors in dairy foods.
> * Specific artificial flavors and colors (except Yellow Dye No. 5).
> * A specific sugar or oil used. (For example a label could say "may contain coconut oil, lard or soybean oil" without specifying which one.)

All statements, claims, drawings and pictures about the product that is put on the label must be accurate (i.e. it can not be misleading).

TABLE I - SAFETY RATING CODES

S	**SAFE** in concentrations allowed and usually used.
C	**CAUTION**. Only consume these ingredients in moderation.
A	**AVOID**. Minimize consumption or avoid completely.
AL	Ingredient has been found to provoke an **ALLERGIC** response in many people.
A-ASM	**ASTHMA** sufferers should avoid this ingredient.
A-ASP	Avoid if **ASPIRIN** sensitive.
A-G	**GOUT** sufferers should avoid.
A-INF C-INF	**INFANTS** should avoid. **INFANTS**: Use only in moderation.
A-PINF	**PREMATURE INFANTS** should avoid.
A-PW C-PW	**PREGNANT WOMEN** should avoid. **PREGNANT WOMEN**: Use only in moderation

CHAPTER 4 - THE ADDITIVES DICTIONARY

A

ACACIA GUM [S/C-PW/AL] (stabilizer/ emulsifier) A vegetable gum, it comes from the Acacia tree. Also called gum Arabic, Gum Acacia or Gum Senegal. GRAS

ACESULFAME-K or -POTASSIUM [C] (sweetener) An artificial sweetener. Trade name of "Sunette." Safety questioned.

ACETATE [S] (flavoring) A salt of acetic acid.

ACETIC ACID [S] (solvent) Occurs naturally in fruits and in a variety of other foods.

ACETONE PEROXIDE [C] (maturing/bleaching) This is acetone with compound containing oxygen added to it.

ACETYLATED MONO- and DIGLYCERIDES [C/AL] (emulsifier). See glycerides.

ADIPIC ACID [S] (acidifier) Also called hexanedoic acid. Used for neutralizing, buffering and flavoring. Contained in beets. GRAS

AGAR-AGAR [S/AL] (stabilizer/thickener) A seaweed, Japanese Isinglass. GRAS

ALBUMIN [S/AL] (emulsifier) A group of water soluble proteins. Usually derived from egg whites (egg albumin or ovalbumin), but also from milk (lactalbumin). [Avoid if allergic to eggs or sensitive to milk.]

ALGAE or KELP [C] (stabilizer) Brown dried seaweed. GRAS

ALGIN or ALGIN GUM [C] (stabilizer) A derivative of alginic acid. Typically a sodium salt of alginic acid. See alginic acid. GRAS

ALGINATES [C] (stabilizer/humectant) From seaweed. Alginic acid that may be used in combination with ammonium, calcium, potassium and sodium. GRAS

ALGINIC ACID [C] (stabilizer/defoaming) Extract of brown dried seaweed. GRAS

ALLURA RED AC [A] See FD&C Colors, Red No. 40.

ALLYL ISOTHIOCYANATE or MUSTARD OIL [S] (flavoring) Found in mustard, onions or horseradish. [Can be used to produce a toxic gas.]

ALPHA-AMYLASE [S/AL] See amylase

ALPHA TOCOPHEROL [S/AL] (vitamin E) See Tocopherol. GRAS

ALUM [C] (firming/thickening) See Aluminum Sulfate. GRAS

ALUMINUM [C] A metal used in compounds, and in food packaging. [It can concentrate in the brain, and has been found there in some victims of dementia-type diseases but its role is not clear.]

ALUMINUM AMMONIUM SULFATE [C] (anticaking/ neutralizer) A sulfuric acid salt. See ammonium and aluminum. GRAS

ALUMINUM CALCIUM SILICATE [C] (anticaking) See silicate and aluminum. GRAS

ALUMINUM HYDROXIDE [C] (alkalizer/leavener) See aluminum. GRAS

ALUMINUM PALMITATE [S] An aluminum salt of palmitic acid. It is safe for its intended use, which is not a direct additive to the food, but is used in food packaging materials. GRAS

ALUMINUM POTASSIUM SULFATE [C] (texturizer/ firming) See potassium and aluminum. GRAS

ALUMINUM SODIUM SULFATE [C] (firming) See sodium and aluminum. A salt of sulfuric acid, sodium and aluminum. GRAS

ALUMINUM SULFATE [C] (firming/thickener) Also called Cake Alum and Patent Alum. A salt of sulfuric acid and aluminum. See aluminum. GRAS

AMMONIATED GLYCYRRHIZIN [C] (flavoring) Also known as licorice. An extract of the Glycyrrhiza Glabra plant also known as Sweet Root. It is black in color. [Excessive consumption is linked to high blood pressure, headaches, and other ailments.] GRAS

AMMONIUM A caustic liquid used in salt compounds. A nitrogen-hydrogen compound. It is manufactured by blowing steam through hot coke.

AMMONIUM ALGINATE [C] (humectant/stabilizer) See alginates and ammonium. GRAS

AMMONIUM BICARBONATE [S] (leavener/alkalizer) Manufactured by blowing carbon dioxide through a concentrated solution of ammonium and water. GRAS

AMMONIUM CARBONATE [S] (leavener) Derived from ammonium bicarbonate. GRAS

AMMONIUM CARRAGEENAN [S/A-PINF] (emulsifier/ stabilizer) See carrageenan. GRAS

AMMONIUM CASEINATE [S/AL] (texturizer) See Casein and ammonium. A protein. It occurs naturally in milk and cheese and does not have to be listed separately on the food label.

AMMONIUM CHLORIDE [S] (dough conditioner) A naturally occurring ammonium salt. See ammonium. GRAS

AMMONIUM CITRATE [S] (firming/flavor enhancer) Occurs naturally in plants and animals. GRAS

AMMONIUM GLUTAMATE [S] (flavor enhancer) A salt of glutamic acid (from vegetable protein). GRAS

AMMONIUM HYDROXIDE [S] (neutralizer /alkalizer) Ammonium bicarbonate dissolved in water. See ammonium. GRAS

AMMONIUM ISOVALERATE or ISOVALERIC ACID or VALERIC ACID [S] (flavoring) Extracted from plants such as the valerian plant, apples, cocoa, coffee and strawberries.

AMMONIUM PHOSPHATE [S] (acidifier/leavener) mono- and dibasic. See phosphate and ammonium. GRAS

AMMONIUM SULFATE [S] (neutralizer/dough conditioner) See ammonium. GRAS

AMMONIUM SULFIDE [S] (seasoning) A synthetic spice.

AMMONIUM SULFITE [C/A-ASM/AL] (preservative) A sulfiting agent. See sulfites.

AMYLASE [S/AL] (bread improver) An enzyme that converts a starch to sugar. Two types are alpha-amylase (found in plants and animals) and beta-amylase (found in plants). In practice it is taken from hog pancreas and used in flour to convert starch to sugar. The yeast enzymes in dough then convert the sugar to carbon dioxide helping dough rise.

ANNATTO [S] (coloring/flavoring) An extract from a tropical tree.

ARABINOGALACTAN [C/AL] (emulsifier /stabilizer) Extracted from Larch wood (see Larch Gum).

ARACHIS OIL [S] See Peanut oil.

ARTIFICIAL FLAVORINGS [C] Artificial flavorings are usually complex formulations of chemicals. Over 1500 chemicals are in use, and it would be too difficult to evaluate each formulation. However, the amount of chemicals used in artificial flavoring is such a small percentage that the average person consumes less than one ounce a year from all sources.

ARTIFICIAL SWEETENERS [C] Artificial means not found in nature. See acesulfame-K, aspartame and saccharin which are currently the common artificial sweeteners.

ASCORBATES [S/AL] (antioxidant) Calcium and sodium salts of ascorbic acid.

ASCORBIC ACID [S/AL] (acidifier/antioxidant) Vitamin C. It can be obtained from plants or synthesized from glucose. [It is incompatible with certain medications.] GRAS

ASCORBYL PALMITATE [S] (preservative /antioxidant) Salt of ascorbic acid. GRAS

ASPARTAME [C/A-PW] (sweetener) Trade name of "Nutrasweet." A compound formed by combining aspartic acid with phenylalanine (both are amino acids). [It must be avoided by those inflicted with phenylketonuria (PKU) because they cannot tolerate phenylalanine. It lowers urine acidity which reportedly increases urinary tract susceptibility to infection.]

ASPERGILLUS [C/AL] (preservative) A type of fungus that acts as an antibiotic.

AZODICARBONAMIDE [S] (bleaching/ maturing)

B

BAKING POWDER [S/AL] (leavening) A yeast substitute. A powder (generally sodium bicarbonate) and an acid (such as Cream of Tartar) along with a starch filler.

BEET RED [S] (coloring) A vegetable dye from beets.

BENZOATE OF SODA [S/AL] (preservative /antifungal) Also known as sodium benzoate. GRAS

BENZOATES [S/AL] (preservative/antifungal) Salts or esters of benzoic acid.

BENZOIC ACID [S/AL] (preservative/antifungal) An organic acid produced from plants. GRAS

BENZOYL PEROXIDE [S] (bleaching) GRAS

BETA-APO-8'-CAROTENAL [S] (coloring)

BETA-AMYLASE [S/AL] See amylase.

BETA-CAROTENE [S] (coloring/nutritional) Found in plants and some animal tissue. Vitamin A is derived from it. It produces a natural yellow coloring. GRAS

BHA [A] (antioxidant/preservative) see Butylated hydroxyanisole. GRAS

BHT [A] (antioxidant/preservative) see Butylated hydroxytoluene. GRAS

BICARBONATE OF SODA [S] (alkalizer/leavener) See Sodium Bicarbonate. GRAS

BIOTIN or VITAMIN H [S] (nutritional) A Vitamin B factor. Found in small amounts in every living cell, and in large amounts in milk and yeast. GRAS

BIOFLAVONOIDS or VITAMIN P COMPLEX [S] (nutritional) Derived from the pulp of citrus fruits.

BIXIN [S] (coloring) Derived from Annatto, it causes a yellow color.

BLUE No. 1 and BLUE No. 2 [A] See FD&C Colors.

BRILLIANT BLUE [A] See FD&C Colors

BROMATES [C] A salt of bromic acid.

BROMELAIN or BROMELIN [S] (meat tenderizer) An enzyme derived from pineapples.

BROMIC ACID [C] An acid of bromine, a chemical element. Bromine is obtained from salt brines as a byproduct in the salt industry.

BROMINATED VEGETABLE OIL or BVO [C/AL] (emulsifier) Bromine combined with vegetable oil. [linked to many ailments]

BUTTER FAT [S] (texturizer/emulsifier) An oily fat obtained from mammal's milk. [caution is advised from the standpoint of excessive consumption of saturated fats if it is one of the first three ingredients]

BUTTERMILK SOLIDS [S/AL] If listed on label, product may contain CASEIN, LACTALBUMIN or WHEY without being listed separately.

BUTYLATED HYDROXYANISOLE or BHA [A/AL] (preservative/ antioxidant) [Can effect liver and kidneys. However, some studies have shown it to have a beneficial effect of reducing the occurrence of tumors and enhancing fetal survival. More study is required.] GRAS

BUTYLATED HYDROXYTOLUENE or BHT [A/AL] (preservative/ antioxidant) [Can effect liver and kidneys. However, some studies have shown it to have a beneficial effect of reducing the occurrence of tumors and enhancing fetal survival. More study is required.] GRAS

1,3 BUTYLENE GLYCOL [A] (humectant/solvent) A petroleum product. [Can adversely affect nervous system. Can damage kidneys.]

BUTYL LACTATE [S/AL] (flavoring) See lactic acid. GRAS

BUTYLPARABEN [C/AL] (preservative/antifungal) Ester of butyl alcohol and parahydroxybenzoic acid.

BUTYRIC ACID [S] (flavoring) See Ethyl Butyrate. GRAS

BVO [C/AL] See brominated vegetable oil.

C

CAFFEINE [C/A-PW] (flavoring) A strong stimulant it occurs naturally in coffee, tea, cocoa, kola beans, chocolate. [Affects nervous system, pulse rate, behavioral changes] GRAS

CALCIFEROL or VITAMIN D$_2$ (nutritional) See Vitamins.

CALCIUM [S] A mineral required by the body for nutrition.

CALCIUM ACETATE [S] (sequestrant) From limes. See acetic acid. GRAS

CALCIUM ALGINATE [C] (texturizer/stabilizer) See alginates. GRAS

CALCIUM ASCORBATE [S/AL] (antioxidant/ preservative) Produced from ascorbic acid and calcium carbonate. See ascorbates. GRAS

CALCIUM BROMATE [C] (maturing/dough conditioner) A combination of the minerals calcium and bromine. The baking process converts the bromate to a bromide which becomes much safer.

CALCIUM CARBONATE [S] (alkalizer/yeast food/ flavor carrier) A chalk that occurs naturally in coral, limestone and marble. [Can cause constipation.] GRAS

CALCIUM CARRAGEENAN [S/C-PW/A-PINF] (emulsifier/ stabilizer) See carrageenan. GRAS

CALCIUM CASEINATE [S/AL] (texturizer) Made from milk protein. See casein. GRAS

CALCIUM CHLORIDE [S] (firming) [Can cause stomach disturbance.] GRAS

CALCIUM CITRATE [S/AL] (buffer/dough improver) Made from citrus fruit. See citric acid. GRAS

CALCIUM DIOXIDE [S] (bleaching) See Calcium Peroxide.

CALCIUM DISODIUM EDTA [S/AL] (preservative/ sequestrant)

CALCIUM FUMARATE [S/AL] (acidifier) A calcium salt of fumaric acid. See fumarate.

CALCIUM GLUCONATE [S/AL] (firming/buffer/ sequestrant) See gluconate. [May cause disturbance of cardiac and gastrointestinal systems] GRAS

CALCIUM HEXAMETAPHOSPHATE [S] (texturizer/ sequestrant/ emulsifier) See phosphates. GRAS

CALCIUM HYDROXIDE [S] (firming) A strong alkali. [Caustic] GRAS

CALCIUM IODATE [C] (dough conditioner/ nutrient) Source of iodine. GRAS

CALCIUM LACTATE [S/AL] (firming/buffer/yeast food) Made from milk or corn. GRAS

CALCIUM OXIDE or QUICKLIME [S] (alkalizer/ yeast food/dough conditioner) Produced by heating limestone to a high temperature. [Caustic] GRAS

CALCIUM PANTOTHENATE [S] (nutritional) A salt of pantothenic acid. A B-Complex vitamin. See Pantothenic Acid. GRAS

CALCIUM PEROXIDE [S] (bleaching) Derived from interaction between sodium peroxide and calcium salts. [Skin irritant]

CALCIUM PHOSPHATE [S] (yeast food/bleaching/ dough conditioner/ anticaking/texturizer) Includes mono- di- and tribasic formulations. See phosphates. GRAS

CALCIUM PHYTATE [S] (sequestering) GRAS

CALCIUM PROPIONATE [S/AL] (antifungal) See propionates. GRAS

CALCIUM SILICATE or OKENITE [S] (anticaking) A salt of calcium and silicic acid. See silicates. GRAS

CALCIUM SORBATE [S] (preservative/antifungal) Calcium salt of sorbic acid. See sorbic acid. GRAS

CALCIUM STEARATE [S] (anticaking) Made from peanuts, soy or corn. See stearic acid. [A saturated fat, it can elevate cholesterol level. Exercise caution if it is one of the first three ingredients on the label.] GRAS

CALCIUM STEAROYL-2 LACTYLATE [S/AL] (stabilizer/ dough conditioner/ emulsifier) Manufactured from lactic and stearic acids. See stearoyls.

CALCIUM SULFATE or PLASTER OF PARIS [S] (yeast food/ maturing/ dough conditioner/ alkalizer) A calcium salt of sulfuric acid. A naturally occurring mineral, a type of gypsum. [May cause constipation, intestinal obstructions.] GRAS

CALCIUM SUPEROXIDE [S] (bleaching) See calcium peroxide.

CANOLA OIL [S] An unsaturated oil from Canola, a variety of rapeseed plant developed to have low erucic acid. Erucic acid was found to be a health problem.

CANTHAXANTHIN [S] (coloring) A pink coloring extract derived from trout, salmon, crustaceans, tropical birds or mushrooms.

CAPRYLIC ACID [S] (flavoring) Found in milk or palm oil. Made by oxidation of octyl alcohol.

CARAMEL [S] (flavoring/coloring) Produced by heating carbohydrates (e.g. burnt sugar). Trace elements of alkali metals may also be added. GRAS

CARBOXYMETHYL CELLULOSE or CMC [S] (stabilizer) A cellulose gum. Made from cotton or wood processed with vinegar. GRAS

CARMINIC ACID or NATURAL RED No. 4 [C/AL] (coloring) Derived from the scales of an insect.

CARNAUBA WAX [C] (texturizer) From the leaves of the Brazilian wax tree. Used to put a glaze on candy, etc. GRAS

CAROB BEAN GUM [S] See Locust Bean Gum. GRAS

CARBON DIOXIDE [S] (carbonation gas) GRAS

CARBONATE [S] An ester or salt of carbonic acid.

CARBONIC ACID [S] A solution of carbon dioxide in water. GRAS

CAROTENE [S] See Beta Carotene. GRAS

CARRAGEENAN [S/C-PW/A-PINF] (emulsifier/ stabilizer) Made from the sea weed known as Irish Moss. GRAS

CASEIN [S/AL] (texturizer) A protein from milk. [Will cause allergic reactions to those sensitive to milk.] GRAS

CASEINATES [S/AL] (texturizer) Caseins dissolved in alkaline solutions and combined with the appropriate compounds to produce ammonium, calcium, magnesium, potassium or sodium caseinate. See casein. GRAS

CELLULOSE GUM [S] (stabilizer) Fibrous substance consisting chiefly of cell walls from plants. Refer to carboxymethyl cellulose for example. GRAS

CHLORINE [S] (maturing/ bleaching) An element that exists as a gas or in combination with other substances. It is very reactive. [Lethal as a free gas if inhaled.]

CHLORINE DIOXIDE [S] (maturing/bleaching) [Corrosive and lethal as a gas.]

CHLOROPENTAFLUORETHANE [S] (pressurizer/ propellant) It is being phased out because of the damage it causes to the ozone layer in the upper atmosphere.

CHOLINE BITARTRATE [S] (nutritional) Found as a thick syrupy liquid in most animal tissue. A part of the Vitamin B-Complex. GRAS

CHOLINE CHLORIDE [S] (nutritional) A Vitamin B-Complex with the same effects as Choline Bitartrate. GRAS

CINNAMALDEHYDE or CINNAMIC ALDEHYDE [A/AL] (flavoring) Derived from a wood fungus. [Irritating to skin and gastrointestinal tract. A common allergen] GRAS

CITRIC ACID or ETHYL CITRATE or TRIETHYL CITRATE [S/AL] (acidifier/ sequestrant/ preservative/ flavoring) Made from fermentation of citric fruits. [Can erode tooth enamel]. GRAS

CITRUS RED No. 2 [A] (coloring) See FD & C Colors.

CLOVE BUD EXTRACT or OIL [S] (flavoring) Extracted from tropical tree's dried flower buds (clove buds) [Can cause intestinal upset.] GRAS

CLOVE BUD OLEORESIN [S] (seasoning) An extract of the resin of the clove bud tree.

CMC [S] See carboxymethyl cellulose. GRAS

COCONUT OIL [S] (solvent/emulsifier) A highly saturated fat from coconut kernels. A lard substitute. [Should be avoided in excess by those with high cholesterol.] GRAS

COPPER [S] (nutritional) Copper is an essential mineral. It may be found in combination with acids as copper carbonate, gluconate, hydroxide or sulfate. [Ingesting copper (or cupric) sulfate can cause vomiting.] GRAS

CORNSTARCH [S/AL] (texturizer/humectant) A starch flour from ground Indian Corn. See starch GRAS

CORN SYRUP [S/AL] (sweetener/ thickener/ humectant) Also called dextrose. Made form cornstarch. GRAS

COTTONSEED OIL [S/AL] (emulsifier) Extracted from seeds of various plants.

CREAM OF TARTAR [S] See potassium acid tartrate. GRAS

CUPROUS IODIDE [C] (bread improver) See iodine. GRAS

CURCUMIN [C] (coloring) Derived from turmeric. See turmeric.

CYANOCOBALAMIN [S] See Vitamin B_{12}.

CYSTEINE [S] (nutritional) An essential amino acid derived from hair. GRAS

D

7-DEHYDROCHOLESTEROL [S] See Vitamin D_3.

DEXTRIN [S/AL] (defoaming/ flavor carrier) Also called starch gum or British gum. Made from grain starch. GRAS

DEXTROSE [S/AL] See corn syrup. GRAS

DIACETYL [S] (flavor carrier) Prepared by fermentation of glucose. Can be found naturally in many foods. GRAS

DIACETYL TARTARIC ACID ESTERS OF MONO- AND DIGLYC-ERIDES [S/AL] (emulsifier) See tartaric acid, glycerides and glycerin. GRAS

DIGLYCERIDES [S/AL] (emulsifier) Esters of glycerin. GRAS

DILAURYL THIODIPROPIONATE [S] (antioxidant) Found in fats and oils. GRAS

DILL OIL [S] (flavoring) From the crushed seeds of the Dill herb. GRAS

DIMETHYL POLYSILOXANE [S] (defoaming) Also called Antifoam A and Dimethicone. A type of silicone.

DIOCTYL SODIUM SULFOSUCCINATE or DSS [C] (solvent/ stabilizer)

DISODIUM EDTA [S/AL] (sequestrant/ preservative) See EDTA.

DISODIUM GUANYLATE or GMP [S/A-G/AL] (flavor enhancer) Isolated from some types of mushrooms.

DISODIUM INOSINATE or IMP [S/A-G/AL] (flavor enhancer) Prepared from meat extract or dried sardines.

DISODIUM PHOSPHATE [S] (emulsifier/ sequestrant/ buffer) A salt of phosphoric acid. See phosphate. GRAS

DISODIUM PYROPHOSPHATE [S] (emulsifier/ texturizer) See Sodium Pyrophosphate. GRAS

DSS [C] See dioctyl sodium sulfosuccinate.

DULSE [S/C-PW] (flavoring) An extract from red seaweed. GRAS

E

EDTA or ETHYLENEDIAMINE TETRAACETIC ACID [S/AL] (antioxidant/ sequestrant)

ERGOCALCIFEROL [S] (nutritional) See Vitamin D_2

ERGOSTEROL [S] (nutritional) A steroid found in yeast and mold. It is converted to Vitamin D upon exposure to ultraviolet light.

ERYTHROBIC or ISOASCORBIC ACID [S/AL] (antioxidant) Made from corn. GRAS

ERYTHROSINE or RED No. 3 [A] See FD&C colors.

ETHOXYLATED MONO- AND DIGLYCERIDES [S/AL] (emulsifier/dough conditioner) See glycerides.

ETHYL ACETATE [S] (flavoring) Found naturally in various fruits. GRAS

ETHYL BUTYRATE or BUTYRIC ACID [S] (flavoring) Also called pineapple oil. Found naturally in apples and strawberries. Has a pineapple odor when put in an alcohol solution. GRAS

ETHYL CITRATE [S] See Citric Acid.

ETHYL CELLULOSE [S] (binder/ filler) A cellulose gum. Made from wood pulp or cotton.

ETHYL FORMATE [S] (antifungal/flavoring) Found naturally in coffee and apples. GRAS

ETHYL HEPTANOATE [S] (flavoring) Synthetically produced, it has a wine-like odor and fruity taste.

ETHYL LACTATE [S/AL] (flavoring) Lactic acid and ethanol. See lactic acid. GRAS

ETHYL MALTOL [S] (flavor enhancer) Made from maltol but much stronger.

ETHYL METHYLPHENYLGLYCIDATE [C] (flavoring) A synthetic berry flavoring also called strawberry aldehyde.

ETHYL PROPIONATE or PROPIONIC ACID [S/AL] (preservative) See propionic acid. GRAS

ETHYL VANILLIN [S] (flavoring) An artificial vanilla flavor. GRAS

F

FAST GREEN [A/AL] See FD&C Colors, Green No. 3

FD & C COLORS (Food Drug and Cosmetic Colors) Colors allowed in foods by the FDA and are identified by numbers instead of formulas.

BLUE NO. 1 or BRILLIANT BLUE. [A] Coal tar derivative.

BLUE NO. 2 or INDIGOTINE [A] Coal tar derivative.

CITRUS RED NO. 2 [A]

GREEN NO. 3 or FAST GREEN [A/AL]

LAKES (Colors combined with aluminum or calcium lakes)

RED NO. 3 or ERYTHROSINE. [A] Coal tar derivative. Contains iodine. [may interfere with brain nerve impulses]

RED NO. 3 ALUMINUM LAKE [A] Red No.3 Aluminum salt.

RED NO. 40 or ALLURA RED AC [A]

YELLOW NO. 5 or TARTRAZINE [C/A-ASP/AL] Coal tar derivative. [found to be dangerous to aspirin sensitive people]

YELLOW NO. 5 ALUMINUM LAKE [C/A-ASP/AL] See Yellow No. 5.

YELLOW NO. 6 or MONOAZO or SUNSET YELLOW FCF [A/AL]

YELLOW NO. 6 ALUMINUM LAKE [A] See Yellow No. 6.

FENUGREEK EXTRACT and OLEORESINS [S] (flavoring) From the seed of the Fenugreek herb (also called Greek Hay). Used to make curry. GRAS

FERRIC or FERROUS or IRON COMPOUNDS [S] (nutritional) Iron is an essential mineral and is often combined with other compounds such as ferric ammonium citrate, ferric phosphate, ferric pyrophosphate, ferric sodium pyrophosphate, ferrous citrate, ferrous lactate, and ferrous sulfate, which are all GRAS.

FERROUS GLUCONATE [S] (flavoring/coloring) Gluconic acid combined with iron. [Can cause digestive disturbance.] GRAS

FISH OIL [S] (emulsifier/solvent) From fish or marine mammals. Consumption reduces risk of coronary disease. GRAS

FOLIC ACID [S] (nutritional) Part of Vitamin B complex. Found in green leaves, kidney, liver and mushrooms.

FORMIC ACID [S] (flavoring) Found in fruits.

FRUCTOSE [S/AL] (sweetener) A naturally occurring sugar in honey and fruit. Avoid large amounts.

FUMARIC ACID [S/AL] (acidifier) Derived from plants directly or synthesized from malic acid. GRAS

FURCELLERAN [C/AL] (thickener/emulsifier/ stabilizer) A Red Seaweed extract.

G

GARLIC EXTRACT and OIL [S] (flavoring) From the garlic plant. Has been found to control intestinal worms. GRAS

GELATIN [S] (stabilizer/thickener) A protein obtained from animal bones, tendons or skin boiled in water. GRAS

GLUCOAMYLASE [S] (catalyst) An enzyme used in food processing to break down sugar.

GLUCONATES [S/AL] (sequestrant) Derived from glucose. GRAS

GLUCONIC ACID [S] (sequestrant/nutritional) Made from corn. GRAS

GLUCONO DELTA-LACTONE [S/AL] (leavener/ acidifier)

GLUCOSE [S/AL] (sweetener/thickener) Derived from fruits and plants.

GLUTAMATES [S/AL] (flavor enhancer) Salts of glutamic acid. GRAS

GLUTAMIC ACID and GLUTAMIC ACID HYDROCHLORATE [S/C-INF/AL] (flavor enhancer/salt substitute) An amino acid. Made from vegetable protein. GRAS

GLUTEN [S/AL] (emulsifier) See wheat gluten.

GLYCERIDE [S/AL] An ester of glycerin.

GLYCERIN or GLYCEROL [S/AL] (solvent/ humectant) A soap manufacturing byproduct. Prepared by hydrolysis of oils and fats. GRAS

GLYCEROL ESTERS of WOOD ROSIN [S] (stabilizer) From refined wood rosin and glycerin.

GLYCERYL ABIETATE [C] (plasticizer/ thickener)˙ Also called ester gum. From pine rosin.

GLYCERYL LACTO-OLEATE and LACTO-PALMITATE of FATTY ACIDS [S] (emulsifier) Its concentration must be compatible with lactic acid concentrates. GRAS

GLYCEROL MONOSTEARATE [S] (emulsifier/ dispersant) A mixture of two glycerines.

GLYCERYL TRIACETATE or TRIACETIN [S] (fixative) Produced by adding acetate to glycerin. GRAS

GLYCINE or AMINOACETIC ACID [S] (nutritional) A nonessential amino acid.

GLYCYRRHIZIN [C] See ammoniated glycyrrhizin. GRAS

GMP [S/A-G/AL] (flavor enhancer) see disodium guanylate.

GUAR GUM [S/C-PW/AL] (thickener/stabilizer) From Guar flour which comes from Guar (a plant from India) seed tissue grindings. GRAS

GUM ARABIC or ACACIA GUM or GUM SENEGAL [S/C-PW/AL] See Acacia gum. GRAS

GUM GHATTI or GHATTI GUM or INDIAN GUM [C/AL] (emulsifier) From the stem of a plant found in Ceylon or India. GRAS

GUM GUAIAC [A/AL] (antioxidant) From guaiacum tree wood. GRAS

GUM KARAYA or GUM KADAYA or STERCULIA GUM [C/AL] (emulsifier) From Sterculia tree, native to India. GRAS

GUM TRAGACANTH [C/AL] (thickener/stabilizer) From plant found in Iran and Syria. GRAS

GURU or KOLA NUT EXTRACT [S] (flavoring) See Kola Nut. GRAS

GREEN No. 3 [A/AL] See FD&C Colors.

H

HELIUM [S] A gas used as a pressurizer to propel food. GRAS

HEPTYLPARABEN [C/AL] (preservative) See parabens.

HESPERIDIN or VITAMIN P [S] (nutritional) A bioflavonoid derived from the pulp of citrus fruits.

HIGH-FRUCTOSE CORN SYRUP [S/AL] (sweetener) Corn syrup that has been sweetened by enzyme treatment. GRAS

HIGH PROTEIN FLOUR [S/AL] If on label, the product may contain albumin, casein or gluten without being listed separately.

HPP [C/AL] See Hydrolyzed Plant or Vegetable Protein. GRAS

HYDROGEN PEROXIDE [S] (preservative/bleach) Produced from phosphoric acid and barium peroxide. It readily breaks down into oxygen and water. GRAS

HYDROGENATED VEGETABLE OIL [S/AL] (solidifier/neutralizer) Vegetable oil which is solidified by combining with hydrogen gas at high pressure. [Caution is advised from the standpoint of excessive consumption of saturated fats if it is listed as one of the first three ingredients.]

HYDROLYZED CASEIN or HYDROLYZED MILK PROTEIN [S/C-INF /AL] Liquified casein. See casein.

HYDROLYZED PLANT PROTEIN (HPP) or HYDROLYZED VEGE-TABLE PROTEIN (HVP) [C/AL] (flavor enhancer) Vegetable proteins liquified by acids or enzymes. GRAS

HYDROXYLATED LECITHIN [A/AL] (emulsifier/ antioxidant) Lecithin treated to add an oxygen and hydrogen atom to increase solubility.

HYDROXYPROPYL CELLULOSE and METHYLCELLULOSE [S] See carboxymethyl cellulose.

HVP [C/AL] See Hydrolyzed Plant and Vegetable Protein. GRAS

I

IMP [S/A-G/AL] (flavor enhancer) See disodium inosinate.

INDIA GUM or GUM GHATTI [C/AL] (emulsifier) See Gum Ghatti. GRAS

INDIAN TRAGACANTH [C/AL] (emulsifier) See Karaya Gum. GRAS

INDIGOTINE [A] See Blue No.2

INOSITOL [S] (nutritional) A Vitamin B complex.

INVERT SUGAR [S/AL] (sweetener) Mixture of 50% fructose and 50% glucose. GRAS

IODINE [S] An element needed by the body in trace amounts. Found abundantly in seafood and dairy products.

IONONE [S/AL] (flavoring) Found in Boronia, an Australian shrub.

IRISH MOSS [S/A-PINF] See carrageenan. GRAS

IRON COMPOUNDS See Ferric or Ferrous.

IRRADIATION [S] (preserving process) Food showered with ionized radiation to kill bacteria and mold.

ISOASCORBIC or ERYTHROBIC ACID [S/AL] (antioxidant) Made from corn. GRAS

ISOLATED SOY PROTEIN [S/AL] (protein source) Protein concentrated from soy beans.

ISOPROPYL CITRATE [S/AL] (sequestrant/ antioxidant) GRAS

ISOVALERIC ACID or VALERIC [S] (flavoring) See Ammonium Isovalerate.

K

KARAYA GUM or KADAYA GUM [C/AL] (emulsifier) See Gum Karaya. GRAS

KELP or ALGAE [C] (stabilizer) Brown dried seaweed. GRAS

KOLA or **GURU NUT EXTRACT** [S] (flavorings) From the seeds of certain trees found in tropical climates. GRAS

L

LACTALBUMIN or **LACTALBUMIN PHOSPHATE** [S/AL] See albumin.

LACTIC ACID [S/AL] (fermenting/flavoring) Present in muscle tissue after glucose metabolism and in beer, sour milk, etc. as a result of fermentation. GRAS

LACTIC YEAST [S] Yeast obtained from milk. See yeast.

LACTOSE [S/AL] (humectant/nutritional) A natural milk sugar.

LACTYLIC STEARATE [S/AL] (dough conditioner) An ester of stearic acid. See stearic acid.

LARCH GUM or **ARABINOGALACTAN** [C/AL] (emulsifier/ stabilizer/ thickener/ texturizer) A resin extracted from Larch trees.

LARD [S] (texturizer/emulsifier) Purified fat from a hogs abdomen. [Excessive consumption can produce high LDL cholesterol. Avoid if it is one of the first three ingredients listed.] GRAS

LECITHIN [S/AL] (emulsifier/defoaming/ antioxidant) Isolated from eggs, corn or soybeans. GRAS

LICORICE [C] See Ammoniated Glycyrrhizin. GRAS

LOCUST BEAN GUM or **CAROB BEAN GUM** or **ST.JOHN'S BREAD** [S/AL] (stabilizer/thickener) Extract from carob tree seeds. GRAS

LYSINE or **L-LYSINE** [S] (nutritional) An essential amino acid taken from casein or fibrin. It is needed for growth. GRAS

M

MACE or NUTMEG OIL and OLEORESIN [C/AL] (flavoring) From the Nutmeg plant. Steam distilled from dried nutmeg seeds. [The nutmeg seeds ingested directly can be very toxic.] GRAS

MAGNESIUM [S] A mineral needed in diet. Found in combination with other chemicals.

MAGNESIUM ACETATE [S] (buffer/neutralizer) A salt of acetic acid. GRAS

MAGNESIUM CARBONATE [S] (alkalizer) A salt of carbonic acid. GRAS

MAGNESIUM CHLORIDE [S] (color fixative/ firming/neutralizer) A salt of hydrochloric acid. GRAS

MAGNESIUM FUMARATE [S/AL] (acidifier) A salt of fumaric acid. See fumaric acid.

MAGNESIUM GLUCONATE [S] (buffering/ neutralizing) A magnesium salt of gluconic acid.

MAGNESIUM HYDROXIDE [S] (color fixative/ drying) GRAS

MAGNESIUM OXIDE [S] (neutralizer) Oxidized magnesium. GRAS

MAGNESIUM PHOSPHATE [S] (nutritional) A salt of phosphoric acid. See phosphate. GRAS

MAGNESIUM SILICATE and TRISILICATE [S] (anticaking) A salt of silicic acid. See silicates. GRAS

MAGNESIUM STEARATE [S] (nutritional) A salt of stearic acid. GRAS

MALIC ACID [S/C-INF] (alkalizer) Found in fruits like apples and cherries. GRAS

MALTODEXTRIN [S/AL] (texturizer/flavor enhancer) A sugar produced by starch hydrolysis. It is a combination of dextrin and maltol. GRAS

MALTOL [S/AL] (flavoring) From the bark of Larch trees.

MALTOSE [S/AL] (nutrient/sweetener) A malt sugar derived from extract of barley (malt).

MANGANESE [S] (nutritional) A mineral needed for development of bones. Found in combination with other compounds, such as manganese chloride, glycerophosphate, hypophosphite, and sulfate. GRAS

MANNITOL [S/AL] (texturizer/sweetener/ anticaking) Found in plants. Mostly taken from seaweed. GRAS

MENADIONE or VITAMIN K$_3$ [S] (nutritional/ preservative) It has the properties of Vitamin K but is produced synthetically.

METHIONINE [S] (nutritional/flavoring) An essential amino acid. Found in protein. GRAS

METHYLCELLULOSE [S] (stabilizer/thickener/ emulsifier) From wood pulp or cotton. GRAS

METHYL ETHYL CELLULOSE [S] (foaming/ emulsifier) Made from wood pulp or cotton.

METHYLPARABEN [C/AL] (preservative) See parabens. GRAS

METHYL SILICONE or METHYL POLYSILICONE [S] (preservative) See silicones.

MILK DERIVATIVES, PROTEINS or SOLIDS [S/AL] If any of these are listed on the label, the product may contain casein, lactalbumin or whey without being listed separately.

MODIFIED STARCH [C/AL] (thickener) Starch modified by various chemicals to change its properties and make it more digestible. Attention should be paid to which chemicals are used to modify it. They may be listed as separate ingredients.

MONO- and DI- GLYCERIDES [S/AL] (emulsifier) Esters of glycerin. GRAS

MONOAMMONIUM GLUTAMATE [S/A-INF/AL] (flavor enhancer) Ester of glutamic acid. GRAS

MONOAZO [A/AL] See FD&C Colors, Yellow No.6.

MONOPOTASSIUM GLUTAMATE [S/A-INF/AL] (flavor enhancer) A salt of glutamic acid. See glutamates. GRAS

MONOSODIUM GLUTAMATE or MSG [C/A-INF/AL] (flavor enhancer) Found naturally in seaweed, sugar beets and soybeans. [Connected with "Chinese restaurant syndrome", where MSG is heavily used, of headaches, numbness and chest pains after eating.] GRAS

MONOSODIUM PHOSPHATE [S] (emulsifier) Derived from edible fats. A salt of phosphoric acid. See phosphate. GRAS

MONOSODIUM PHOSPHATE DERIVATIVES OF MONO- and DI-GLYCERIDES [S/AL] (emulsifier) Derived from edible fats. GRAS

MSG [C/A-INF/AL] See monosodium glutamate. GRAS

MUSTARD [S] (flavoring) The yellow and white mustard is from the seeds of the Brassica alba variety of mustard plant. The black, brown and red mustard is from the Brassica Nigra variety. GRAS

MUSTARD OIL or ALLYL ISOTHIOCYANATE [S] (flavoring) Found in mustard, onions or horseradish. [Can be used to produce a toxic gas.]

N

NATURAL RED No. 4 [C/AL] (coloring) See Carminic Acid.

NIACIN or NIACINAMIDE or VITAMIN B₃ [S] (nutritional) Found in whole cereals, legumes, meat, etc. GRAS

NICOTINAMIDE or NICOTINIC ACID [S] See niacin.

NITRATES [A] (color fixative/preservative) Prevents botulism. Also found in vegetables that are heavily fertilized with nitrates. [Linked to cancer]

NITRITES [A] (color fixative/preservative) Nitrates exposed to air form nitrites. Prevents botulism. [Linked to cancer, but effects appear to be counteracted by taking with food high in vitamin C (eg. grapefruit juice).]

NITROSYL CHLORIDE [S] (bleaching) Very corrosive as a gas.

NITROUS OXIDE [S] (pressurizer) Also known as laughing gas. GRAS

NUTMEG or MACE OIL and OLEORESIN [C/AL] (flavoring) From the Nutmeg plant. Steam distilled from dried nutmeg seeds. [The nutmeg seeds ingested directly can be very toxic.] GRAS

O

OAT GUM [S/AL] (thickener/stabilizer/ antioxidant) A plant extract. GRAS

OIL of CLOVE, GARLIC, MACE, NUTMEG See Clove Bud Oil, Garlic Oil, Mace Oil, and Nutmeg Oil.

OLEIC ACID [S] (defoaming/binding) From plant or animal fats and oils. GRAS

OLIVE OIL [S] (solvent/emulsifier) A mono unsaturated fat from olives. Reported to reduce cholesterol in blood.

OVALBUMIN [S/AL] See albumin.

OXYSTEARIN [C] (defoaming/ anti-crystallization) Mixture of stearic and other fatty acids with glycerides. GRAS

P

PALMITIC ACID [S] (seasoning) A solid organic acid mixture found naturally in many animal fats, coffee, tea, celery seed and palm oil (from which it is largely obtained).

PALM OIL [S] (solvent) From Oil Palm Trees native to Malaysia and Central Africa. [Caution is advised from the standpoint of excessive consumption of saturated fats and cholesterol, if it is one of the first three ingredients listed.]

PANTOTHENIC ACID or VITAMIN B$_5$ (nutritional) Found in rice bran, molasses, liver. A synthetic form, *d*-pantothenamide is made form yeast, molasses and queen bee jelly.

PAPAIN [S/AL] (meat tenderizer/clarifier) An enzyme from papayas. GRAS

PAPRIKA [C] (flavoring/coloring) In powder and oleoresin form. It is obtained from sweet pepper pods. Reddish-orange in color.

PARABENS [C/AL] (preservatives/antifungal) Esters of parahydroxybenzoic acid.

PEANUT OIL or ARACHIS OIL [S] (solvent/ flavoring) An oil pressed from peanuts. GRAS

PECTIN [S] (stabilizer/thickener/binder/ emulsifier) From stems, roots or fruits of plants. GRAS

PEPSIN [S] (nutritional) An enzyme that aids in digestion. Obtained from the hog's stomach glands.

PHOSPHATES [S] (emulsifier/texturizer/ sequestering) Salts or esters of phosphoric acid. Phosphate is required by the body for many functions, but too much can cause kidney damage.

PHOSPHORIC ACID [S] (sequestering/acidulant) Made from phosphate rock. GRAS

PHOSPHORUS [S] A mineral element. See phosphates and phosphoric acid.

POLYDEXTROSE [S/AL] (filler) Used to reduce calories. Made from corn.

POLYGLYCEROL and POLYGLYCEROL ESTERS [S] (plasticizer/emulsifier/dispersant/ humectant) Prepared from edible fats in plants and animals.

POLYOXYETHYLENE SORBITAN MONOOLEATE or POLYSORBATE 80 [S/AL] (emulsifier/defoaming/ flavor carrier) A condensate of sorbitol and oleic acid.

POLYOXYETHYLENE SORBITAN MONOSTEARATE or POLYSORBATE 60 [S/AL] (emulsifier/ defoaming/flavor carrier) A condensate of sorbitol and stearic acid.

POLYOXYETHYLENE SORBITAN TRISTEARATE or POLYSORBATE 65 [S/AL] (emulsifier/ defoaming/flavor dispersant) A condensate of sorbitol and stearic acid.

POLYSORBATES [S/AL] (emulsifier/stabilizer) Compounds of fatty acids and polyoxyethylene sorbitan.

POTASSIUM [S] An essential mineral (from the alkali metal group) found in chemical compounds.

POTASSIUM ACID TARTRATE or CREAM OF TARTAR [S] (acidifier/buffer) Salt of tartaric acid. GRAS

POTASSIUM ALGINATE [C] (stabilizer/ humectant) A salt of potassium and alginic acid. See alginates. GRAS

POTASSIUM BICARBONATE [S] (leavener) A salt of carbonic acid. GRAS

POTASSIUM BISULFITE [C/A-ASM/AL] (preservative/ antioxidant) A sulfiting agent. See sulfites. GRAS

POTASSIUM BROMATE [C] (bread improver) A salt of bromic acid. The process of baking converts the bromate to a bromide, which improves its safety.

POTASSIUM CARBONATE [S] (alkalizer) A salt of potassium also known as salt of Tartar or Pearl Ash. GRAS

POTASSIUM CARRAGEENAN [S/A-PINF] (emulsifier/ stabilizer) See carrageenan. GRAS

POTASSIUM CASEINATE [C] (texturizer) A potassium salt of milk protein. See caseins.

POTASSIUM CHLORIDE [S] (yeast food/salt substitute) A potassium salt of hydrochloric acid. GRAS

POTASSIUM CITRATE [S/AL] (buffer/alkalizer) Citric acid potassium salt. GRAS

POTASSIUM FUMARATE [S/AL] (acidifier) A potassium salt of fumaric acid.

POTASSIUM GLUCONATE [S/AL] (buffer) A gluconic acid potassium salt. GRAS

POTASSIUM GLUTAMATE [S] See glutamates. GRAS

POTASSIUM HYDROXIDE [S] (alkalizer/solvent) Also called Caustic Potash. Prepared by electrolysis of potassium chloride.

POTASSIUM IODATE [C] (nutritional) Source of iodine. A potassium salt of iodic acid. GRAS

POTASSIUM IODIDE [C] (nutritional) Source of iodine. Added to table salt. GRAS

POTASSIUM METABISULFITE or **POTASSIUM PYROSULFITE** [C/A-ASM/AL] (preservative/ antioxidant) A sulfiting agent. See sulfites. GRAS

POTASSIUM NITRATE or **SALT PETER** [A] (color fixative) See nitrates.

POTASSIUM NITRITE [A] (color fixative) See nitrites.

POTASSIUM PHOSPHATE [S] (yeast food) Mono- Di- and Tribasic. A salt of phosphoric acid. See phosphate. GRAS

POTASSIUM SORBATE [S] (preservative/ antifungal) A potassium salt of sorbic acid. GRAS

POTASSIUM SULFITE [C/A-ASM/AL] (preservative) A sulfiting agent. See sulfites.

PROPIONATES [S/AL] Salts or esters of propionic acid. GRAS

PROPIONIC ACID [S/AL] (preservative) Obtained from certain plant fermentation. Found naturally in apples, tea, and wood pulp. [May cause migraine headaches for those susceptible.] GRAS

PROPYL GALLATE [A/AL] (antioxidant) FDA limits amount to .02% or less. [may cause stomach irritation.] GRAS

PROPYLENE GLYCOL and PROPYLENE GLYCOL MONO- AND DIESTERS [S/AL] (emulsifier/ humectant/ color fixative) GRAS

PROPYLENE GLYCOL ALGINATE [S/AL] (stabilizer/ filler/defoaming) An ester of alginic acid (from seaweed) combined with propylene glycol. GRAS

PROPYLENE GLYCOL MONOSTEARATE [S/AL] (emulsifier/dough conditioner) An ester of stearic acid with propylene glycol. GRAS

PROPYLPARABEN [C/A-PW/AL] (preservative/ antifungal) Esters of parahydroxybenzoic acid. GRAS

PYRIDOXINE or PYRIDOXINE HYDROCHLORIDE or VITAMIN B$_6$ **[S]** (nutritional) From yeast and grains. GRAS

Q

QUICKLIME [S] See calcium oxide. GRAS

QUILLAIA or QUILLAJA EXTRACT [C] (flavoring) Also called Soap Bark, Quillay Bark, Panama Bark, China Bark Extract. From the bark of the South American Quillaja saponaria tree.

QUININE and QUININE HYDROCHLORIDE and QUININE SULFATE [C/A-PW] (flavoring) Extracted from the bark of the cinchona tree, native to South America. [Overdose causes nausea and vision and hearing disturbances.]

R

RAPESEED OIL [C] (solvent/emulsifier) Derived from an annual herb called a turnip-like. See also Canola Oil. [Contains erucic acid, which was found to be a possible source of heart problems. Be especially concerned if it is one of the first three ingredients listed on the label.]

RED No.3 and RED No. 3 ALUMINUM LAKE [A] See FD&C Colors.

RED No. 40 [A] See FD&C Colors.

RENNET or RENNIN [S] (texturizer) A milk curdling agent. An enzyme taken from stomach lining of calves. GRAS

RIBOFLAVIN or VITAMIN B2 [S] (nutritional) Also called lactoflavin. Every living cell (plant or animal) contains a minute amount. GRAS

RIBOFLAVIN-5'-PHOSPHATE [S] See Riboflavin. A more soluble form of riboflavin. GRAS

S

SACCHARIN [A] (artificial sweetener) A carcinogen. A warning label is required.

ST. JOHN'S BREAD GUM [S/AL] See locust bean gum. GRAS

SAFFLOWER OIL [S] (solvent/emulsifier) From the seed of a European thistle plant. Claim have been made that its consumption can prevent or reduce fat build-up in the arteries.

SAFROLE [A] (flavoring) Derived from natural oils found in nutmeg, ylang-ylang and star anise plants. [Banned in 1960 as a beverage flavoring.]

SALT [S] (preservative/seasoning) Popular name for sodium chloride. Technically a salt is any chemical compound of an acid and a metal. GRAS

SAPP [S] See sodium acid pyrophosphate. GRAS

SHELLAC or CONFECTIONER'S GLAZE [C/AL] (texturizer) A resin produced from certain tree feeding insect excretions.

SILICATES [S] (anticaking) Esters of salts of silicic acid. Found abundantly in earth's mineral crust.

SILICON [S] A non-metallic element found largely in silica form.

SILICONES [S] (anticaking) Oils or resins derived from silica.

SILICON DIOXIDE [S] (anticaking/defoaming) Also called silica. Found in many forms: sand, quartz, etc. GRAS

SMOKE FLAVORING [C] (flavoring) Also called liquid smoke and char-smoke flavor. Maple and hickory are preferred. Made from many sources of condensed and purified wood smoke. [Smoke processed foods have been associated with increased cancer risk. Smoke flavored foods contain some of the same ingredients that are found in smoked foods. Although smoke flavoring should be less risky than smoked foods, caution is still advised.]

SODIUM [S] A metal common in combination with other chemicals, like sodium chloride (table salt).

SODIUM ACETATE [S] (preservative/alkalizer) An acetic acid sodium salt. GRAS

SODIUM ACID PYROPHOSPHATE or SAP [S] (buffer/sequestrant) Sodium combined with phosphoric acid. GRAS

SODIUM ALGINATES [C] (stabilizer) Sodium salt of alginic acid (extracted from seaweed). GRAS

SODIUM ALUMINATE and SODIUM ALUMINOSILICATE [C] (alkalizer/coloring) Used to manufacture lake colors. (See FD&C Colors) GRAS

SODIUM ALUMINUM PHOSPHATE [C] (buffer) A salt of phosphoric acid. [Caution urged because of body's tendency to accumulate aluminum.] GRAS

SODIUM ALUMINUM SULFATE [C] (bleaching) A salt of sulfuric acid. [Caution urged because of body's tendency to accumulate aluminum.]

SODIUM ASCORBATE [S/AL] (antioxidant) Vitamin C (ascorbic acid) with sodium. GRAS

SODIUM BENZOATE [S/AL] (preservative/ flavoring) Sodium salt of benzoic acid. GRAS

SODIUM BICARBONATE or BICARBONATE OF SODA [S] (alkalizer/leavener) Also known as baking soda. Prepared by combining soda ash with carbon dioxide. GRAS

SODIUM BISULFITE [C/A-ASM/AL] (preservative) Sulfiting agent. See sulfites. GRAS

SODIUM CALCIUM ALUMINOSILICATE [S] (anticaking) See silicates. GRAS

SODIUM CARBONATE or SODA ASH [S] (humectant/ neutralizer) Found naturally in ores and lake or sea brines. GRAS

SODIUM CARBOXYMETHYL CELLULOSE [S] (stabilizer/thickener) See carboxymethyl cellulose. GRAS

SODIUM CARRAGEENAN [S/A-PINF] (stabilizer/ emulsifier) See carrageenan. GRAS

SODIUM CASEINATE [S/AL] (texturizer) A milk protein (casein) partly neutralized with sodium hydroxide. GRAS

SODIUM CHLORIDE [S] (preservative/seasoning) Common table salt. [Avoid if blood pressure is high.] GRAS

SODIUM CITRATE [S/AL] (emulsifier/buffer/ sequestrant) Sodium salt of citric acid. GRAS

SODIUM DIACETATE [S] (preservative) Combination of sodium acetate and acetic acid. Vinegar smell. GRAS

SODIUM ERYTHORBATE or SODIUM ISOASCORBATE [S/AL] (preservative/antioxidant) See erythorbic acid. GRAS

SODIUM FERROCYANIDE or YELLOW PRUSSIATE OF SODA [S] (anticaking) Produced by combining sodium carbonate, iron and other organic compounds with heat.

SODIUM FUMARATE [S/AL] (acidifier) Sodium salt of fumaric acid.

SODIUM GLUCONATE [S/AL] (sequestrant) A gluconic acid sodium salt. GRAS

SODIUM HEXAMETAPHOSPHATE and **SODIUM METAPHOSPHATE** [S] (emulsifier/sequestrant/ texturizer) Also called Graham's Salt. See phosphate. GRAS

SODIUM HYDROXIDE [S] (alkalizer/emulsifier) Also called Caustic Soda and Soda Lye. GRAS

SODIUM ISOASCORBATE or **SODIUM ERYTHORBATE** [S/AL] (preservative) See erythorbic acid. GRAS

SODIUM LAURYL SULFATE [S] (surfactant/ emulsifier) A sulfation of lauryl alcohol neutralized by sodium carbonate.

SODIUM METABISULFITE [C/A-ASM/AL] (preservative/antibrowning) A sulfiting agent. See sulfites. GRAS

SODIUM METAPHOSPHATE [S] (dough conditioner) Related to sodium hexametaphosphate. Also known as Graham's Salt. See phosphate. GRAS

SODIUM NITRATE [A] See nitrates.

SODIUM NITRITE [A] See nitrites.

SODIUM PECTINATE [S] (stabilizer/thickener) See pectin. GRAS

SODIUM PHOSPHOALUMINATE [C]. [S for packaging] (leavening) A salt of phosphoric acid. See aluminum and phosphate. GRAS for packaging.

SODIUM POTASSIUM TARTRATE or **ROCHELLE SALT** [S] A sodium-potassium salt of tartaric acid. GRAS

SODIUM PROPIONATE [S/AL] (humectant/ preservative) A salt of propionic acid. See propionates. GRAS

SODIUM PYROPHOSPHATE or **SODIUM IRON PYROPHOSPHATE** [S] (emulsifier/sequestrant/ texturizer) See phosphate. GRAS

SODIUM RIBOFLAVIN PHOSPHATE [S] Vitamin B with sodium. See Riboflavin. GRAS

SODIUM SESQUICARBONATE [S] (neutralizer) Produced from sodium carbonate. Also called Lye. GRAS

SODIUM SILICATE [S] (anticaking) Also called Water Glass and Soluble Glass. See silicate. GRAS

SODIUM SILICO ALUMINATE [C] (anticaking) See silicates and aluminum.

SODIUM SORBATE [S] (preservative) See sorbic acid. Sodium salt of sorbic acid. GRAS

SODIUM STEAROYL FUMARATE [S/AL] (dough conditioner) A salt of fumaric acid. See stearoyls.

SODIUM STEAROYL-2-LACTYLATE [S/AL] (emulsifier/plasticizer) Sodium salt of lactic and fatty acids. See lactic acid.

SODIUM SULFATE [C] (preservative) Naturally occurring mineral also called Salt Cake.

SODIUM SULFITE [C/A-ASM/AL] (preservative) A sulfiting agent. See sulfites. GRAS

SODIUM SULFO-ACETATE DERIVATIVES [S/AL] (emulsifier)

SODIUM TARTRATE [S] (sequestrant/stabilizer) A sodium salt of tartaric acid. GRAS

SODIUM THIOSULFATE [S] (antioxidant/ stabilizer/neutralizer) GRAS

SODIUM TRIPOLYPHOSPHATE or **STPP** [S] (texturizer/sequestrant) A salt of phosphoric acid. See phosphate. [Excess use can deplete body of calcium.] GRAS

SORBIC ACID [S] (antifungal) Obtained from the berries of the mountain ash tree. GRAS

SORBITAN MONOPALMITATE [S/AL] (emulsifier/flavor carrier) See sorbitol.

SORBITAN MONOSTEARATE [S/AL] (emulsifier/ flavor carrier/defoaming) See sorbitol.

SORBITAN TRISTEARATE [S/AL] (emulsifier) See sorbitol.

SORBITOL [S/AL] (texturizer/humectant/ anticaking/sequestrant) An alcohol found in berries of the mountain ash tree and also in other non-citrus fruits and in seaweed. [Foods with likely consumption in excess of 50 grams per day require warning labels of its laxative effect.] GRAS

SOUR CREAM SOLIDS [S/AL] If listed on label, product may contain casein, lactalbumin or whey without being listed separately.

SOYBEAN OIL [S] (defoaming/solvent/ emulsifier) Oil from Soybeans.

SOY CONCENTRATES and ISOLATES [S/AL] (defoaming/ flavor enhancer) Flour and oils extracted from soybeans. Soy sauce is an hydrolysis product of soybeans. GRAS

SOY SAUCE [S] (flavoring) A hydrolyzed and fermented soybean oil Aspergillus molds are used in the fermentation process. GRAS

STANNOUS CHLORIDE or TIN DICHLORIDE [S] (antioxidant) GRAS

STARCH [S] (thickener) Found naturally in plants. GRAS

STEARIC ACID [S] (flavoring/texturizer) Found naturally in vegetable and animal oils and fats. [A saturated fat, it can elevate cholesterol level. Exercise caution if it is one of the first three ingredients on the label.] GRAS

STERCULIA GUM [C/AL] (emulsifier) See Gum Karaya. GRAS

STEAROYL PROPYLENE GLYCOL HYDROGEN SUCCINATE or SUCCINISTEARIN [S/AL] (emulsifier) See succinic acid.

STEAROYLS [S/AL] (dough conditioner) Made from soy, corn, peanuts, milk.

STEARYL CITRATE [S/AL] (antioxidant/sequestrant) Ester of citric acid and stearyl alcohol. GRAS

STEARYL LACTATE or STEARYL-2-LACTYLIC ACID [S] (emulsifier) Found in animal fats and vegetable oils.

STPP or SODIUM TRIPOLYPHOSPHATE [S] (texturizer/ sequestrant) See phosphate. [excess use can deplete body of calcium.] GRAS

SUCCINIC ACID [S] (buffer/neutralizer) Derived from acetic acid. GRAS

SUCCINISTEARIN or STEAROYL PROPYLENE GLYCOL HYDRO-GEN SUCCINATE [S/AL] (emulsifier) See succinic acid.

SUCCINYLATED MONO- AND DIGLYCERIDES [S] (dough conditioner/surfactant) See succinic acid and glycerides.

SUCROSE or SUGAR [S] (sweetener) From plants like sugar cane and sugar beets. [Excessive consumption stimulates body fat production.] GRAS

SUCROSE POLYESTER or SPE or OLESTRA [S] (texturizer/ emulsifier) A fat substitute. It is passed through the body undigested, saving consumption of fat calories. It is made by reacting fatty acids with table sugar (sucrose)

SULFITES [C/A-ASM/AL] (preservatives/ antioxidants) Combined with ammonium, potassium or sodium to produce sulfiting agents. Use of sulfites must be clearly stated on label aside from the ingredients list. [Reactions include severe asthma attacks in which several deaths have occurred.]

SULFUR DIOXIDE [C/A-ASM/AL] (preservative/ antioxidant/antibrowning/bleaching) Produced as a gas by burning sulfur in air. A sulfiting agent. See sulfites. GRAS

SULFURIC ACID [S] (acidifier) Also called Oil of Vitriol. GRAS

SUNFLOWER SEED OIL [S] (solvent/emulsifier) From sunflower seeds.

SUNSET YELLOW [A/AL] See FD&C Colors, Yellow No. 6.

T

TALC [C] (anticaking) Also called French Chalk. It is finely powdered magnesium silicate. [A lung irritant if inhaled. Associated with cancer.] GRAS for packaging.

TALLOW [S] (defoaming) Also called suet. Fat from sheep or cattle. [Caution is advised from the standpoint of excessive consumption of saturated fats if it is listed as one of the first three ingredients on the label.]

TANNIC ACID or TANNIN [C] (clarifier/ refining/flavoring) Obtained from bark and fruit of plants. Found in tea and coffee. GRAS

TARTARIC ACID [S] (flavoring/acidifier/ sequestrant/emulsifier) Obtained from fruits. A byproduct of wine making. GRAS

TARTRAZINE [C/A-ASP/AL] (coloring) See FD&C Colors, Yellow No. 5.

TBHQ or TERTIARY BUTYLHYDROQUINONE [C] (antioxidant) It contains butane derived from petroleum. [Limited to .02% of fat content. Toxic in larger doses.]

TEXTURIZED VEGETABLE PROTEIN or TVP [S] (filler) A protein source where protein has been isolated from soy and then combined with other additives to resemble meat in appearance, taste and nutrition.

THBP OR 2-4-5 TRIHYDROXYBUTROPHENONE [C] (antioxidant) Limited to .02% of fat content and .005% of total.

THIAMINE HYDROCHLORIDE or THIAMINE NITRATE or VITAMIN B₁ [S] (nutritional) GRAS

TITANIUM DIOXIDE [S] (coloring) A naturally occurring mineral. A white pigment. May not exceed 1% of weight content.

TOCOPHEROL or ALPHA TOCOPHEROL or VITAMIN E [S/AL] (nutritional) Obtained by vacuum distillation of vegetable oils. GRAS

TRIACETIN or GLYCERYL TRIACETATE [S] (fixative) Produced by adding acetate to glycerin. GRAS

TRAGACANTH GUM [C/AL] See Gum Tragacanth.

TRICALCIUM SILICATE [S] (anticaking) Salt of silicic acid. See silicate. GRAS

TRIETHYL CITRATE [S/AL] See Citric Acid.

TURMERIC and TURMERIC OLEORESIN [C] (flavoring/ seasoning/ coloring) The processed root of an East Indian herb. For the oleoresin, the plant's resin is extracted with an alcohol based solvent. GRAS

V

VALERIC ACID or ISOVALERIC ACID [S] (flavoring) See Ammonium Isovalerate.

VANILLA [S] (flavoring) Extracted from the fruit of the vanilla plant (of Mexico) just before ripening. GRAS

VANILLIN [S] (flavoring) Made from vanilla and potato parings. Also made from clove oil synthetically. GRAS

VEGETABLE GUMS [C/AL] (emulsifier/thickener) Derived from various plants. Used with preservatives.

VEGETABLE OIL [S] (texturizer/emulsifier) Obtained from plants (e.g. peanut oil, olive oil, etc.)

VITAMINS [S] (nutritional) Catalysts used by the body for various functions.

VITAMIN A - ACETATE and PALMITATE. An essential vitamin for growth and development. Helps fight infection [prolonged overdoses of 100,000 units/day was shown to be toxic.] Recommended daily allowance is 4,500 units/day for adults and 3,000 for children. GRAS

VITAMIN B_1 - THIAMINE (HYDROCHLORIDE or NITRATE). Helps body regulate energy requirements and is needed by the nervous system.

VITAMIN B_2 - RIBOFLAVIN. Needed for healthy skin, eyes and respiration. GRAS

VITAMIN B_3 - NIACIN. Deficiencies affect the mind, and cause pellagra.

VITAMIN B_5 - PANTOTHENIC ACID. Needed for metabolism of proteins and fats.

VITAMIN B_6 - PYRIDOXINE HYDROCHLORIDE. Needed for metabolism of fats and amino acids (proteins).

VITAMIN B_{12} - CYANOCOBALAMIN. Needed for healthy blood. Produced by intestinal microorganisms. GRAS

VITAMIN C - ASCORBIC ACID. Required by teeth, bones and blood vessels. GRAS

VITAMIN D₂ - **CALCIFEROL** or **ERGOCALCIFEROL.** Speeds body's production of calcium. GRAS

VITAMIN D₃ - **ACTIVATED 7-DEHYDROCHOLESTEROL.** Speeds body's production of calcium. GRAS

VITAMIN E - **TOCOPHEROL.** Obtained by vacuum distillation of vegetable oils. Helps form blood cells. GRAS

VITAMIN H - **BIOTIN.** Vital to growth and blood circulation. GRAS

VITAMIN P - **BIOFLAVONOID** or **HESPERIDIN.** Needed to maintain healthy blood vessels.

W

WHEAT GLUTEN [S/AL] (emulsifier) A protein mixture taken from wheat flour by washing the starch out of dough.

WHEY or **WHEY PROTEIN** [S/AL] (binder/filler) Also called milk or lactic serum; it is the serum left after removing casein and fat from milk. Whey is used to make cheese. [Will cause allergic reactions to those sensitive to milk.] GRAS

X

XANTHAN GUM [S/AL] (emulsifier/thickener/ stabilizer) Also called sugar gum. Produced by fermentation of a carbohydrate (like corn) with a culture of xanthomas campestris (a bacteria).

XYLITOL [A] (artificial sweetener) Made from pulp waste.

Y

YEAST [S] (catalyst) An enzyme producing fungus. There are different types of yeast. Ordinary Yeast. Brewers Yeast. Bakers Yeast. GRAS

YEAST-MALT SPROUT EXTRACT [S/AL] (flavor enhancer) See yeast.

YELLOW PRUSSIATE OF SODA or SODIUM FERROCYANIDE [S] (anticaking) Produced by combining sodium carbonate, iron and other organic compounds with heat.

YELLOW No. 5 and YELLOW No. 5 ALUMINUM LAKE [C/A-ASP/AL] See FD&C Colors.

YELLOW No. 6 and YELLOW No. 6 ALUMINUM LAKE [A/AL] See FD&C Colors.

Z

ZINC [S] (nutritional) A metal needed by the body in trace amounts. Combined with other compounds, as Zinc Acetate, Zinc Gluconate, etc.

APPENDIX I. REFERENCES

1. Freydberg, Nicholas, Ph.D., and Gortner, Willis, Ph.D., "The Food Additives Book," Bantam Books, New York, 1982

2. Jacobson, Michael, Ph.D., "The Complete Eater's Digest and Nutrition Scoreboard," Revised & Updated, New York, Anchor Press Doubleday, New York, 1985.

3. Winter, Ruth, "A Consumers Dictionary of Food Additives," Third Revised Edition, Crown Publishers, New York, 1989.

APPENDIX II. GLOSSARY

Acetylated: An organic compound heated with acetyl chloride or acetic anhydride. It is used to produce a coating on food that seals in moisture.

Acid: A substance that forms free hydrogen ions (H+) when dissolved in water. The solution will have a pH of less than 7.0. The lower the number, the higher the acidity.

Acidifier: A substance that increases acidity or reduces alkalinity.

Acidulant: An acidifier that adds a tangy or tart taste to a product.

Alkali: A substance that forms free hydroxyl ions (OH-) when dissolved in water. The solution will have a pH of greater than 7.0. The higher the number, the higher the alkalinity.

Alkalizer: A substance that increases alkalinity or reduces acidity.

Allergen: A substance causing allergic reactions to those susceptible.

Allergy: A hypersensitivity to specific substances (allergens) due to an altered immune response.

Amino Acids: Building blocks of proteins. There are 22 known amino acids. Eight of these are called essential because they are needed for health and are not produced in the body.

Antibrowning: Prevents browning due to oxidation.

Anticaking: Prevents powders from hardening.

Anticrystallization: Prevents forming of crystals that can cloud up certain products.

Antifungal: Inhibits growth of rope, mold, and fungus.

Antimicrobial: Prevents the growth of bacteria and fungus.

Antioxidant: A substance that prevents food from oxidizing, which would change its color and/or taste.

Artificial: A substance that is not found in nature (even though its components can be natural ingredients).

Bases: Alkalis (See Alkalis).

Binder: A substance that holds ingredients together.

Bleaching Agent: A whitener. It accelerates the natural aging process of flour (improves baking).

Bran: The outer shell of cereal grain. It is indigestible, but provides bulk and fiber in the diet.

Bread Improver: Oxidizing substances which mature or "improve" the quality of bread.

Buffer: A substance that helps regulate the pH (acidity or alkalinity) to a prescribed level.

Butters: Substances which are soft but solid at room temperature but melt at body temperature.

Carbohydrate: A major class of food mostly from plants (like sugars, starches and cellulose) and composed of carbon, hydrogen and oxygen.

Carcinogen: A substance that can produce cancer.

Catalyst: A substance that starts or speeds up chemical reactions without itself changing.

Chelator: A substance that binds to and precipitates out metals. Often used to eliminate trace metals. Also called a chelating agent or sequestering Agent.

Clarifier: Used for removing small particles from liquids.

Cocarcinogen: A substance that increases the potency of carcinogens.

Color Fixative: It keeps the natural color fixed.

Cupric or Cuprous: Compounds containing copper.

Defoaming (or Antifoaming) Agent: Prevents or lessens foam in liquids.

Dispersant: An agent that forms and/or stabilizes the dispersion of one substance throughout another substance.

Distilled: A substance separated from another by boiling off its vapor and then condensing the vapor separately.

Dough Conditioner: Makes dough easier to manage.

Emulsifier: It stabilizes mixtures (like oil and water) and keeps them at a uniform consistency.

Enriched: Products that have had nutrients added to replace some of those that were lost in processing.

Enzymes: A type of protein that acts as a catalyst. They form in living cells and produced unique chemical reactions.

Ester: An alcohol combined with an acid. In the process of forming an ester, water is eliminated and hydrogen radicals of the acid are replaced by an alcohol radical.

Fat: A concentrated source of food energy. A carrier for essential unsaturated fatty acid and fat soluble vitamins.

FDA: The Food and Drug Administration, founded in 1906 to regulate food, drugs and cosmetics.

Fermentation: A breakdown of fats, sugars or starches to other components (e.g. a carbohydrate breaks down into carbon dioxide and alcohol). It is usually accomplished with the aid of bacteria or yeast.

Ferric or Ferrous: Compounds containing iron.

Fiber: Also called "bulk." It is an indigestible carbohydrate. Although not digestible, the fiber has a cleansing effect in the bowels during the digestive process.

Filler: Used for replacing certain ingredients with less costly substitutes.

Fixative: An agent that slows down vaporization of flavor or odor, making it last longer.

Firming Agent: It maintains the firmness and texture of a product.

Flavor Carrier or Flavor Dispersant: It dissolves a flavor and disperses it throughout a product.

Flavor Enhancer: Improves or brings out the natural flavor.

Foam Inhibitor or Antifoamer or Defoamer: Reduces the amount of foam produced in drinks or during food processing.

Foaming Agent: Helps whipped topping peak.

Fortified: Food that has had nutrients added to it in excess of those found naturally.

Gel: A homogeneous semi-solid substance with jelly-like elastic properties.

GRAS: Generally Recognized As Safe. A list established by Congress in 1958 and partially re-evaluated in 1980. It exempted from pre-market clearance those substances that had been in use a long time, and identified their permitted concentrations.

Hormones: A gland produced chemical that is secreted into the blood stream and that affects the function of other cells.

Humectant: A substance used to preserve moisture content.

Hydrolyzed: The process by which a compound is decomposed into simpler components by use of water. (It reacts to the water ions $H+$ and $OH-$.)

Imitation: Flavorings or scentings containing any non-natural materials are called imitation. Also a standardized food (e.g. mayonnaise) must be called imitation (or by some other name) if it fails to use the standard ingredients.

Laxative: A substance that promotes or causes bowel movements.

Leavener: It lightens the texture by releasing gases.

Maturing Agent: Speeds up the aging process in order to make more manageable dough.

Metabolism: Molecular synthesis, decomposition or introconversion by live organisms.

Minerals: The non-organic atoms, such as metals like iron, copper, zinc, cobalt and calcium, that are needed by the body in small amounts.

Mold: A downy or furry growth on organic matter caused by fungus.

Mutagenic: Can cause mutations.

Mutation: A change in a gene's characteristics that will be reproduced in later cell divisions.

Neutralizer: A buffer that eliminates both acidity and alkalinity, keeping the product neutral.

Nucleic Acid: The chemically encode molecules that store genetic information for an organism. Two types of nucleic acids are deoxyribonucleic acid (DNA) and ribonucleic acid (RNA).

Oleoresin: A plant extract of its essential oil and resin by use of alcohol, which acts as a solvent.

Pasteurization: The process of killing microorganisms in food by heat treating it.

pH: An acidity scale from 1 to 14. Neutral is 7.0. Less that 7.0 is acid, and greater than 7.0 is alkaline.

ppm: Parts per million.

Precipitate: To separate out as a solid a substance that was in suspension or solution with another liquid substance.

Preservatives: Antispoilants. Helps prevent food from going bad, or changing taste and color.

Propellant: A gas under pressure, used to expel a containers contents.

Protein: An essential constituent of living cells, composed of nitrogen, carbon, oxygen and hydrogen and some minerals. It is a combination of amino acids.

Radical: A group of atoms (two or more) that acts as a single atom and goes through a reaction unchanged.

Rancidity: The odor of spoiling food caused primarily by the oxidation of unsaturated fats.

Reducing Agent: Any agent that concentrates the volume of a substance or de-oxidizes it. It also prevents oxidation.

Reduction: The reverse of oxidation.

Refining Agent: It purifies products by removing trace elements.

Resins: A substance formed from secretions of plants. It is usually hardened and brittle and translucent or transparent.

Rope: Gelatinous threads formed in loaves of bread caused by a spore forming bacteria.

Salt: Combination of a metal with an acid.

Sequestrant: An agent that binds and precipitates out metal ions that would affect the products appearance, flavor and texture.

Solidifier: Helps to solidify the product.

Solvent: A liquid that can dissolve one or more substances.

Stabilizer: A substance that maintains consistency or texture and gives a product body.

Surfactant: A wetting agent. It allows water to spread out by lowering its surface tension.

Teratogen: A substance that causes developmental damage in a fetus. It is termed teratogenic.

Texturizer: Used to improve texture.

Thickener: For adding body and thickness to the texture.

Toxic: Poisonous. A substance that can damage or degrade body cells.

Vitamins: A complex substance found in foods and essential in small amounts for certain body functions.

Waxes: A plastic substance that is insoluble in water; easily molded at room temperature; and melts at about 148 degrees F. It is obtained from insects, animals, petroleum and plants.

Yeast Food: It speeds up the fermentation process.

APPENDIX III. QUICK GUIDE

SUBSTANCES TO AVOID (EVERYONE):

BHA (Butylated Hydroxyanisole)
BHT (Butylated Hydroxytoluene)
Blue No. 1 (Brilliant Blue)
Blue No. 2 (Indigotine)
1,3 Butylene Glycol
Citrus Red No. 2
Gum Guaiac
Hydroxylated Lecithin
Nitrates
Nitrites
Potassium Nitrate or Salt Peter
Potassium Nitrite
Propyl Gallate
Red No. 3 (Erythrosine)
Red No. 3 Aluminum Lake
Red No. 40 (Allura Red AC)
Saccharin
Safrole
Sodium Nitrate
Sodium Nitrite
Xylitol
Yellow No. 6 (Monoazo or Sunset Yellow)
Yellow No. 6 Aluminum Lake

ADDITIONAL SUBSTANCES TO AVOID IF ASTHMATIC:

Ammonium Sulfite
Potassium Bisulfite
Potassium Metabisulfite or Potassium Pyrosulfite

Potassium Sulfite
Sodium Bisulfite
Sodium Metabisulfite
Sodium Sulfite
Sulfites (Any Sulfiting Agents)
Sulfur Dioxide

ADDITIONAL SUBSTANCES TO AVOID IF ASPIRIN SENSITIVE:

Yellow No. 5 (Tartrazine)
Yellow No. 5 Aluminum Lake

ADDITIONAL SUBSTANCES TO AVOID IF AFFLICTED WITH GOUT:

Disodium Guanylate (GMP)
Disodium Inosinate (IMP)

ADDITIONAL SUBSTANCES TO AVOID FOR INFANTS:

Carrageenan or Irish Moss (if premature)
Monopotassium Glutamate
Monosodium Glutamate (MSG)

ADDITIONAL SUBSTANCES TO AVOID FOR PREGNANT WOMEN:

Aspartame
Caffeine
Propylparaben
Quinine

SUBSTANCES TO BE CAUTIOUS OF (EVERYONE):

Acesulfame-K or -Potassium
Acetone Peroxide
Acetylated Mono- and Diglycerides
Alginates
Algae or Kelp
Algin and Alginic Acid
Aluminum Compounds
Ammoniated Glycyrrhizin (licorice)
Ammonium Alginate
Ammonium Sulfite
Arabinogalactan
Artificial Flavorings
Artificial Sweeteners
Aspartame
Aspergillus
Bromates
Bromic Acid
Brominated Vegetable Oil (BVO)
Caffeine
Calcium Alginate
Calcium Bromate
Calcium Iodate
Carminic Acid
Carnauba Wax
Cuprous Iodide
Curcumin
Dioctyl Sodium Sulfosuccinate (DSS)
FD & C Yellow No. 5 or Tartrazine
FD & C Yellow No. 5 Aluminum Lake
Furcelleran
Gum Ghatti or Ghatti Gum or Indian Gum
Gum Karaya or Karaya Gum or Sterculia Gum

Gum Tragacanth or Tragacanth Gum
Hydrolyzed Plant Protein (HPP)
Hydrolyzed Vegetable Protein (HVP)
Indian Tragacanth
Kelp or Algae
Larch Gum or Arabinogalactan
Mace (Nutmeg)
Methylparaben
Modified Starch
Monosodium Glutamate (MSG)
Oxystearin
Parabens
Paprika
Potassium Alginate
Potassium Bisulfite
Potassium Bromate
Potassium Caseinate
Potassium Iodate and Iodide
Potassium Metabisulfite or Potassium Pyrosulfite
Potassium Sulfite
Propylparaben
Quillaia or Quillaja Extract
Quinine
Rapeseed Oil
Shellac
Smoke Flavoring
Sodium Alginates
Sodium Aluminate
Sodium Aluminum Phosphate
Sodium Aluminum Sulfate
Sodium Bisulfite
Sodium Metabisulfite
Sodium Phosphoaluminate